Guide to Revised Higher Biology Essays

by

John Di Mambro
B.Sc., B.A., M.A., Dip.Ed.Tech., C.Biol., M.I.Biol.

© *John Di Mambro, 1993*
ISBN 0 7169 3178 8

ROBERT GIBSON · Publisher
17 Fitzroy Place, Glasgow, G3 7SF.

INTRODUCTION

It is common knowledge among experienced biology teachers that essays for the Higher Grade Examination present particular difficulties for even the best candidates. Such candidates may perform poorly, not through lack of subject knowledge, but by not collating relevant facts and presenting them in essay form. This may be due to lack of practice since teachers may be unable to devote enough time to the issuing and marking of essays.

To satisfy an obvious need, therefore, over fifty essay questions are given along with well-ordered 'Points to Cover' outlining various facts which can be used in answering each question. Some questions are short while others are more highly structured.

While the Revised Higher Biology Syllabus articulates closely with all the aims of the Standard Grade Biology Syllabus, emphasis is given at Higher Grade to developing particular topics, such as physiological control, biochemistry and ecology. These changes have been reflected where appropriate in the essays which follow.

ADVICE TO CANDIDATES

1. Each point to cover commences with a capital letter and related points are indented.

2. In the 'Points to Cover' for each question, points which could be substituted by the use of appropriately labelled diagrams are marked by a bracket running down the left-hand margin.

3. Where a double-space occurs in a list of points, this indicates the start of a new paragraph.

4. The Topic (in capitals) and Sub-Topic (in lower case) from the Scottish Examination Board's syllabus for the Revised Higher Grade Biology have been given for each essay in the Contents page. This should enable you to easily locate an essay for a particular section of the syllabus.

5. As a guide for candidates, two specimen answers based on the relevant list of 'Points to Cover' are included, these are for Essay Questions 1 and 2. For each essay, one specimen answer is purely textual while the other achieves a similar standard with the use of appropriately labelled diagrams.

6. The list of points to be covered form the skeleton of the answers required. It will be for you, the candidate, to 'flesh' these out using properly constructed sentences and paragraphs. Where appropriate show connections between these by means of link words such as 'therefore', 'because', 'in order to', etc.

7. Before starting your answer, make a short plan. Candidates often feel they cannot spare this time, yet experience shows that a few minutes spent here will result in a better constructed essay. Such a plan reduces the risk of omitting relevant facts and ideas or omitting them from opening paragraphs and having to add them in paragraphs where they do not belong.

8. Be careful to pace yourself in this section of the examination. There is a law of diminishing return for the time you spend on a single question — so spread your effort over both questions. If you were to devote all your time to just one essay, even with a perfect answer, you could only score fifteen marks out of thirty. It is easier to pick up more marks by doing both essays.

9. Your information must be relevant. The inclusion of irrelevant information, even if correct, will gain you no marks.

10. The points listed for inclusion in essay answers to the following questions are not the total of all information on each subject and you may be able to add facts and ideas from your own knowledge. If these are correct and relevant, you will be credited accordingly. This is not necessary, however. High scores would be attainable even if some of the points listed here were omitted and a good essay composed from the remainder.

11. A good scientific essay is a combination of knowledge and expression requiring practice. You can hope to improve only by repeated attempts, thus building up your confidence, timing and expertise.

COPYING PROHIBITED

Note: This publication is NOT licensed for copying under the Copyright Licensing Agency's Scheme, to which Robert Gibson & Sons are not party.

All rights reserved. No part of this publication may be reproduced; stored in a retrieval system; or transmitted in any form or by any means — electronic, mechanical, photocopying, or otherwise — without prior permission of the publisher Robert Gibson & Sons, Ltd., 17 Fitzroy Place, Glasgow, G3 7SF.

CONTENTS

Specimen Answers

1. How is the structure of a chloroplast related to its function? 8
   Specimen answers for Question 1 9
2. Discuss an example of how gene action can be controlled in bacteria 11
   Specimen answers for Question 2 12

CELL BIOLOGY

Cell variety in relation to function

3. By means of four examples, show how the differences between the structure of cells in different tissues are related to the different functions of these tissues. 14

Absorption and secretion of materials

4. Give an account of how materials enter and leave cells under the following headings:
   (a) process of diffusion and osmosis;
   (b) role of the cell wall and cell membrane in these processes;
   (c) how the plasma membrane functions in active transport. 16
5. Give an account of the distribution and functions of cell membranes. 18

Photosynthesis

6. Write an essay on photosynthesis under the following headings:
   (a) role of leaf pigments;
   (b) photolysis;
   (c) ecological importance. 19
7. Describe how carbon dioxide is reduced in the Calvin cycle and the possible final photosynthetic products. 21

Energy release

8. Discuss the role of ATP in living matter. 22
9. Write an essay on mitochondria under the following headings:
   (a) relationship between structure and function;
   (b) occurrence of mitochondria in living cells. 23
10. In what ways do the aerobic and anaerobic phases of respiration differ in animal tissue? 24

Synthesis and release of proteins

11. Discuss the variety of proteins and the role they play in cell structure and function. 25
12. Write an essay on the structure of the DNA molecule and how this structure allows replication to take place. 27
13. Write an essay on nucleic acids under the following headings:
    (a) the differences between DNA and RNA;
    (b) the differences between messenger RNA and transfer RNA. 29

14. Give an account of the structure and function of the following sub-cellular components:
    (a) ribosomes;
    (b) Golgi apparatus;
    (c) rough endoplasmic reticulum. 31

Cellular response in defence

15. Write an essay on viral infection of human cells under the following headings:
    (a) natural invasion of human cells;
    (b) natural barriers to viral invasion;
    (c) alteration of host cells. 33
16. Distinguish between the two different types of specific immune response by humans to disease. 35
17. By means of an example you have studied, explain how humans can mobilise a whole-body response to antigenic invasion. 36
18. Outline some of the mechanisms which plants have to avoid invasion by foreign bodies. 37

GENETICS AND EVOLUTION

Variation

19. Discuss how mitosis and meiosis differ. 38
20. Write an account of sex-linked inheritance in humans. 39
21. Describe the two main forms of mutation and how these might arise. 40

Selection

22. A knowledge of genetics contributes to improvements in agriculture. By means of examples you have encountered, discuss this statement under the following headings:
    (a) plant-breeding;
    (b) hybrid vigour in animals. 42
23. Write an essay on evolution under the following headings:
    (a) adaptative radiation;
    (b) natural selection. 44
24. Discuss evolution today by referring to each of the following:
    (a) industrial melanism;
    (b) resistance. 46
25. How may a species become extinct? 48
26. By means of examples, show how the ability to manipulate genes can be useful to humans. 49
27. Describe how a new species might arise. 50

CONTROL OF GROWTH AND DEVELOPMENT

Growth differences between plants and animals
28. Discuss the role of meristems in higher plants. 51
29. Contrast growth and regeneration in angiosperms and mammals. 52

Genetic control
30. By referring to each of the following, explain how the genotype can influence growth and development:
    (a) development of cell;
    (b) development of an organism. 53

Hormonal influences
31. How is human growth affected by the pituitary and thyroid glands? 55
32. Discuss the influence of indoleacetic acid on plant growth with reference to the following:
    (a) effect on cells and organs;
    (b) practical applications. 56

Environmental influences
33. By means of examples, show how the environment can influence growth and development. 58
34. Give an account of mineral nutrition in plants under the following headings:
    (a) function of macroelements;
    (b) experimental evidence for the importance of macroelements. 59
35. How is flowering in plants affected by the daylength? 61

REGULATION IN BIOLOGICAL SYSTEMS

Physiological homeostasis
36. Discuss the role of the skin and hypothalamus in maintaining a constant body temperature. 63
37. By means of one example, explain what is meant by a homeostatic mechanism. 64

Population dynamics
38. Over a long period of time, a population size usually remains stable. Explain why this happens. 66
39. Write an essay on the factors which influence population change under the following headings:
    (a) density dependent factors;
    (b) density independent factors. 67
40. Discuss succession and climax in plant communities. 69
41. Discuss population change under the following headings:
    (a) population monitoring;
    (b) factors governing population size. 70

ADAPTATION

Maintaining a water balance

42. Why is water so important in the biological world? 72
43. How is water-balance maintained in a:
    (a) marine bony fish;
    (b) desert mammal? 73
44. Give an account of how water is taken up by a mature oak tree and its passage through the stem until it is lost as water vapour to the atmosphere. 75
45. What adaptations for water balance are shown by plants living in the following environments:
    (a) harsh desert conditions;
    (b) aquatic conditions? 76

Obtaining food

46. When resources become limited, competition may arise in animal populations. By means of examples, explain how interspecific and intraspecific competition can occur. 78
47. What are the advantages of social behaviour? 80
48. Give an account of different ways in which animals can communicate with each other. 82
49. Discuss the effect of grazing by herbivores with reference to the following:
    (a) maintaining of species diversity;
    (b) ability of plants to tolerate grazing. 84

Coping with dangers

50. Give an account of how animals may have their behaviour modified under the following headings:
    (a) habituation;
    (b) imprinting. 86
51. By means of suitable examples, show how plants and animals cope with danger in their environment. 87

1 HOW IS THE STRUCTURE OF A CHLOROPLAST RELATED TO ITS FUNCTION?

POINTS TO COVER

Large sub-cellular disc-shaped organelle found in green plant cells but not in animal cells.
The number present in a cell varies from tissue to tissue.
Serves as a site of photosynthesis,
    Which it can continue even if isolated from the whole cell.
Entire structure is related to this function.

Surrounded by a double membrane,
    Which allows metabolites of photosynthesis to enter and leave.

Within chloroplast are the grana,
    Each one of which is composed of up to about one hundred layers of membranes called lamellae,
    Presenting a large surface area,
    Important in allowing capture and transfer of energy in the light stage of photosynthesis.
Running between the grana are membranes called intergranar lamellae.
Each granum consists of protein and lipid molecules alternating with photosynthetic pigment molecules,
    Which absorb solar radiation.

Within the chloroplast is a protein containing matrix called the stroma,
    Which surrounds the grana.
The stroma contains the enzymes needed for the dark stage of photosynthesis.
    Also contains starch granules,
        A known product of photosynthesis,
    Can continue its carbon-fixation function even if dissociated from the lamellae,
        Providing it has a supply of carbon dioxide, adenosine triphosphate and reduced hydrogen carrier,
Supporting role of the stroma in the dark stage.

SPECIMEN ANSWERS — QUESTION 1

EXAMPLE 1

Chloroplasts are large disc-shaped organelles found in varying numbers in green plant cells, but not at all in animal cells. The primary function of the chloroplast is photosynthesis, which it can carry out even if isolated from the whole cell. It is reasonable to expect the structure of this organelle to be closely related to its primary function.

All metabolites involved in photosynthesis, such as water, carbon dioxide, oxygen and so on, must be able to gain access or pass out of the chloroplast. This means that the double membrane which surrounds the chloroplast must allow these chemicals entry or exit.

Within the chloroplast are many membranous structures called grana made up of very thin lamellae. These lamellae present a large surface area to hold many chlorophyll molecules, supported by protein and lipid molecules. This in turn encourages optimum absorption of solar radiation to allow the reactions of the light stage of photosynthesis to take place.

The grana are supported by a fluid-filled matrix called the stroma which contains the enzymes needed to operate the dark stage of photosynthesis. The fixation of carbon by these enzymes can take place even in the absence of lamellae providing the raw materials carbon dioxide, adenosine triphosphate and reduced hydrogen carrier are available. The stroma also holds starch granules which are a known end-product of photosynthesis.

EXAMPLE 2 (with a diagram)

Chloroplasts are large disc-shaped organelles found in varying numbers in green plant cells, but not at all in animal cells. The primary function of the chloroplast is photosynthesis, which it can carry out even if isolated from the whole cell. It is reasonable to expect the structure of this organelle to be closely related to its primary function.

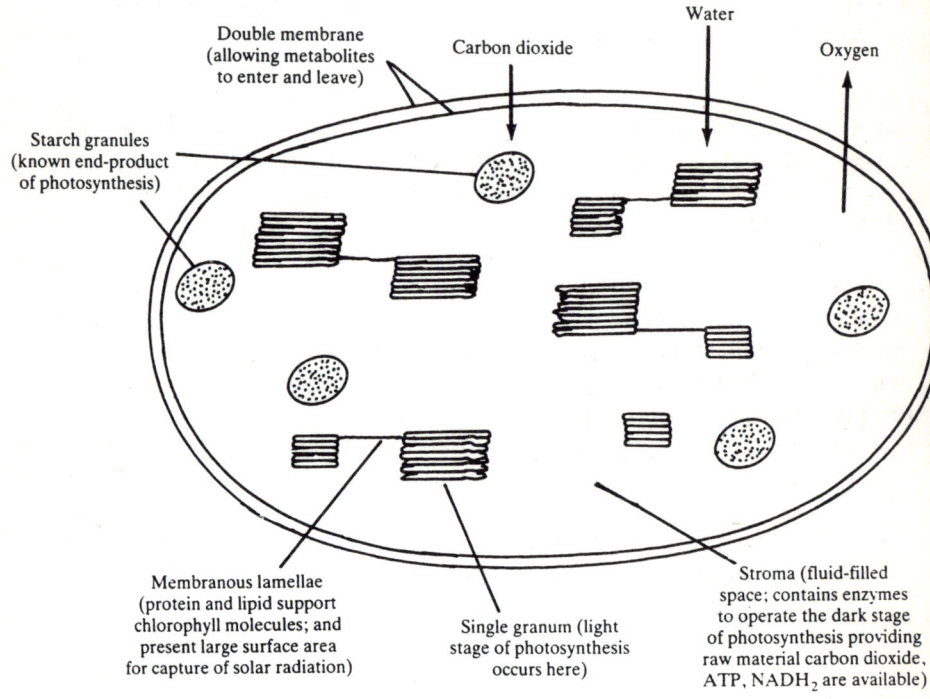

2. DISCUSS AN EXAMPLE OF HOW GENE ACTION CAN BE CONTROLLED IN BACTERIA

POINTS TO COVER

Jacob and Monod proposed a model of gene regulation of enzyme synthesis in *Escherichia coli*,
    A normal part of the human gut microbial population.
The enzyme is called β-galactosidase,
    And breaks down lactose into glucose and galactose.
β-galactosidase synthesis is effected by the activity of a group of structural genes.
⎡The gene cluster which performs such regulatory function is called the lac operon.
⎢In order to avoid wasted production of the enzyme in the absence of its substrate,
⎢   An operator gene switches on or off the structural genes which control the enzyme synthesis.
⎢When lactose is not available to *Escherichia coli*,
⎢   The enzyme is not needed and its level falls.
⎢When lactose is freely available to *Escherichia coli*,
⎢   The enzyme level rises dramatically.
⎢In the absence of lactose, the regulatory gene is constantly being transcribed,
⎢   To be translated into a repressor molecule,
⎢   Which binds to the operator,
⎢   And switches off the structural genes,
⎢   Preventing their transcription.
⎢In the presence of lactose, repressor molecules are inactivated by being bound to the lactose,
⎢   And therefore cannot bind to the operator,
⎢   Or block the transcription of the structural genes,
⎢   And so the synthesis of the enzyme β-galactosidase is switched on.
⎢As long as the inducer lactose is present,
⎢   The synthesis of the enzyme will continue.
⎢When the inducer supply is exhausted,
⎢   Repressor proteins become free again to bind to the operator,
⎣   Effectively switching off the synthesis of the enzyme.

SPECIMEN ANSWERS — QUESTION 2

EXAMPLE 1

Jacob and Monod proposed a model of gene regulation of an enzyme called β-galactosidase which breaks down lactose into glucose and galactose. This enzyme is produced by a bacterium called *Escherichia coli* which is a normal part of the human gut population. The synthesis of β-galactosidase is effected by the activity of a group of structural genes. The gene cluster which performs this regulatory function is called the lac operon. Such regulation avoids wasted production of the enzyme if the substrate is missing. If the substrate lactose is not available, an operator gene switches off the structural genes and the enzyme is not synthesised. If the lactose is available, the structural genes are switched on and the level of β-galactosidase rises dramatically. When lactose is not available, the regulatory gene is constantly being transcribed, to be translated into a repressor molecule which binds to the operator. Thus the structural genes are switched off preventing their transcription. When lactose is available, the repressor molecules are inactivated by being bound to the lactose. Since the repressor molecules cannot now bind to the operator and block the transcription of the structural genes, the synthesis of β-galactosidase is now switched on. So long as the inducer molecule lactose is present, the synthesis of β-galactosidase will continue. When the supply of the inducer molecule is exhausted, repressor proteins become free again to bind to the operator which effectively switches off the synthesis of β-galactosidase.

EXAMPLE 2 (with a diagram)

Jacob and Monod proposed a model of gene regulation of an enzyme called β-galactosidase which breaks down lactose into glucose and galactose. This enzyme is produced by a bacterium called *Escherichia coli* which is a normal part of the human gut population. The synthesis of β-galactosidase is effected by the activity of a group of structural genes.

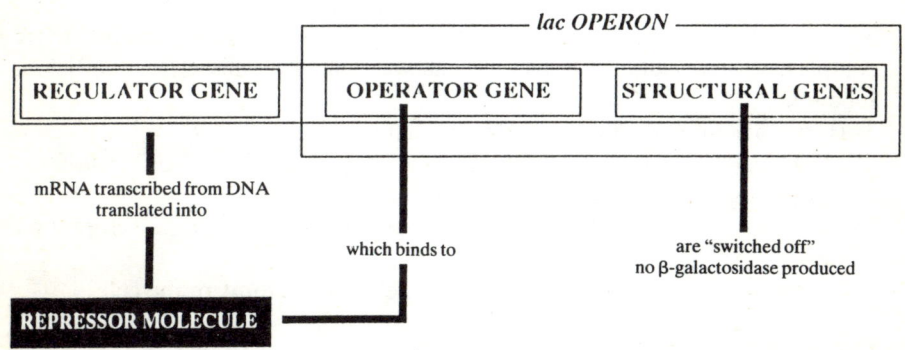

2. LACTOSE PRESENT

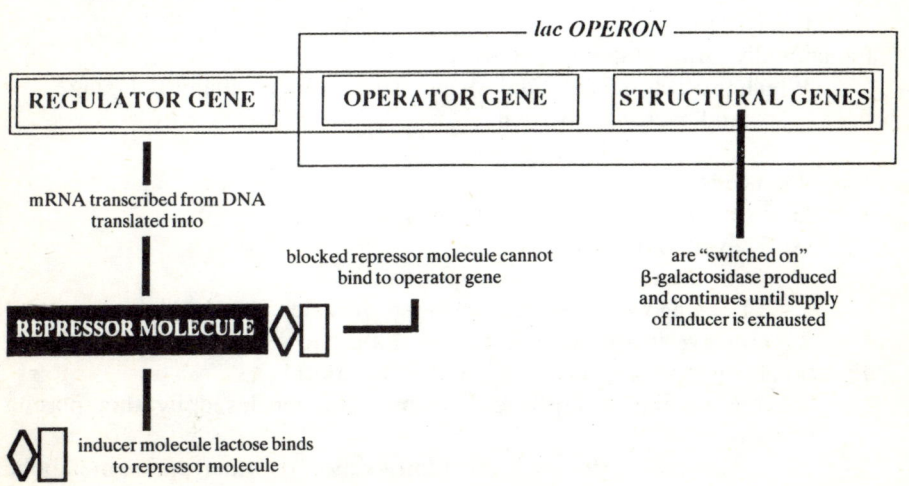

3 **BY MEANS OF FOUR EXAMPLES, SHOW HOW THE DIFFERENCES BETWEEN THE STRUCTURE OF CELLS IN DIFFERENT TISSUES ARE RELATED TO THE DIFFERENT FUNCTIONS OF THESE TISSUES.**

(Four examples are shown, two from animal and two from plant tissues but candidates should choose animal or plant tissues with which they are familiar.)

POINTS TO COVER

When an animal or plant grows, its body structure often becomes more complex as particular groups of cells become specialised for particular functions.
This specialisation is effected by differentiation at the cellular level.
There is usually a very close relationship between structural differentiation and physiological function at the cellular level.
In extreme cases, cells may lose their nuclei or their entire cytoplasmic contents,
    But more often, the alteration is in the basic anatomy of the cell.

Skin epidermal cells
Structurally thickened and flattened,
    And filled with keratin (cornified),
        Which is waterproof and highly resistant to frictional damage,
    Easily shed.
Functionally forms tough outer cover,
    Which protects delicate underlying tissues from mechanical damage,
        And bacterial invasion.

Red blood cells
Structurally circular, biconcave and very small,
    And devoid of nucleus,
        Or other sub-cellular components,
            Leaving room for relatively large volume of haemoglobin.
    Present a very large surface area in relation to volume.
Functionally highly specialised for oxygen transport,
    Having nothing but pigment haemoglobin enclosed by thin plasma membrane,
        And unable to carry out reproduction or other major metabolic activities.
    Very large surface area for gaseous exchange.

Parenchyma cells
Structurally large, vacuolated and often spherically shaped,
    May possess chloroplasts,
    Thin-walled and loosely packed.
Functionally act as packing tissue,
    Often supporting pith and cortex in herbaceous plants,
        By means of turgidity,
    May act as packing in xylem and phloem tissue,
    And may hold food reserves in form of starch.

Xylem vessel cells
Structurally form long tubes,
    Joined end to end,
    When mature, lose their cross-walls,
    Become lignified,
    Eventually die.
Functionally specialised for support and transport,
    Of water and salts,
    In long continuous tubes,
        Without obstruction.

4 GIVE AN ACCOUNT OF HOW MATERIALS ENTER AND LEAVE CELLS UNDER THE FOLLOWING HEADINGS:

(a) processes of diffusion and osmosis;

(b) role of the cell wall and cell membrane in these processes;

(c) how the plasma membrane functions in active transport.

POINTS TO COVER

(a) Processes of diffusion and osmosis

Diffusion is a process by which particles of a substance move from a region of high concentration of that substance to a region of low concentration of that substance (down a concentration gradient),
    Which eventually results in an even distribution of the particles within the containing vessel or available space.
    Occurs much faster in gases than in liquids.
Important as a mechanism for supplying raw materials to metabolising cells,
    And removing products from these cells.
Also important in gaseous exchange.

Osmosis is a special case of diffusion,
    In which water molecules diffuse from an area of high concentration to an area of low concentration across a semi-permeable membrane.
Important in maintaining turgidity.
Also important as contributory factor in setting up a transpiration stream.

(b) Role of the cell wall and cell membrane in these processes

Cell wall is usually permeable to air, water, salts and sugars.
Normally is non-selective over which materials can enter/leave cell.
Presence of cell wall means that plant cells can develop turgidity without bursting,
    Can therefore withstand greater changes in the osmotic environment than animal cells which lack cell walls.

Cell membrane provides a mechanism for exerting control over which materials can enter/leave cell,
    A property called selective permeability.
Water can diffuse through easily while other substances either cannot pass across or do so very slowly,
    Allowing osmosis to take place.

(c) **How the plasma membrane functions in active transport**

Active transport is the movement of materials across a membrane against a concentration gradient,
    And therefore requires the expenditure of energy,
        Supplied in the form of adenosine triphosphate,
    Allowing accumulation, (or maintenance of low concentration) of substances within a cell.

Plasma membrane normally actively transports a particular substance either into or out of cell but not both.
Suggested that within the plasma membrane there are carrier molecules,
    Which react chemically with the substance being actively transported,
    And carry it across the membrane,
    Carrier molecule then becomes free to repeat process.

5. GIVE AN ACCOUNT OF THE DISTRIBUTION AND FUNCTIONS OF CELL MEMBRANES

POINTS TO COVER

Plasma membrane
    Selectively permeable,
    Regulates flow of materials into and out of the cell,
    Phagocytosis allows uptake of large particulate matter,
    Water movement across membrane permits osmosis.

Nuclear membrane
    Allows materials to pass in and out of the nucleus via nuclear pores,
    Extension of endoplasmic reticulum,
    Encloses genetic material.

Golgi membrane
    Formation of vesicles containing carbohydrate-protein molecules.

Lysosomal membrane
    Isolates powerful hydrolytic enzymes,
        Thus preventing autolysis.

Smooth outer mitochondrial membrane
    Allows pyruvic acid to enter and respiratory products to leave.

Mitochondrial cristae
    Convoluted inner membrane provides increased surface area for biochemical reactions.

Endoplasmic reticulum
    Provides a means of intra-cellular transport of cell metabolites,
        Forms compartments within cytoplasm.

Outer chloroplast membrane
    Allows photosynthetic metabolites in and out.

Granar membrane
    Pigments used for photosynthesis located here.

Vacuolar membrane
    Different permeability from plasma membrane's permeability,
    Allows different rates of osmosis, which in turn allows turgidity to develop.

6 WRITE AN ESSAY ON PHOTOSYNTHESIS UNDER THE FOLLOWING HEADINGS:

(a) role of leaf pigments;

(b) photolysis;

(c) ecological importance.

POINTS TO COVER

(a) Role of leaf pigments

Light falling on green leaf
    May pass through (that is, be transmitted),
    Be absorbed,
    Or be reflected from it.
Only light which has been absorbed can be used in photosynthesis.

Chlorophyll a is universal pigment present in photosynthetic plants,
    It is considered essential for photosynthesis,
    For which reason it is called a principal pigment,
    Has important light-capturing role,
        In red and orange wavelengths.

Chloroplasts often contain accessory pigments which can absorb light in other wavelengths,
    And pass on the absorbed energy to chlorophyll a.
Such accessory pigments include chlorophyll b,
    Found particularly in vascular plants and many algae,
    Carotenoids (carotenes and xanthophylls),
        Found in most chloroplasts,
        And absorb wavelengths of light not usable by the chlorophylls.

(b) Photolysis

Biochemical process whereby water molecule is split into hydrogen ions and oxygen,
    The hydrogen combines with an acceptor,
    While the oxygen is released in molecular form.
The energy for the reaction comes from absorbed light.
Provides chemical energy to drive dark stage.
Occurs in grana of chloroplast.

(c) **Ecological importance**

Photosynthesis is the primary method of bringing energy into ecosystems,
> To be trapped in the form of energy rich chemicals,
> Which can be utilised directly or indirectly by other non-photosynthetic life forms.

All natural ecosystems have a photosynthetic component as producer.

Photosynthesis releases free molecular oxygen to be used by animals,
> While removing carbon dioxide breathed out by such animals.

7 DESCRIBE HOW CARBON DIOXIDE IS REDUCED IN THE CALVIN CYCLE AND THE POSSIBLE FINAL PHOTOSYNTHETIC PRODUCTS

POINTS TO COVER

Occurs in the dark stage of photosynthesis
    Whose rate is strongly influenced by temperature.

Carbon dioxide diffuses into chloroplast,
    And is reduced to form carbohydrate in a series of cyclical reactions collectively called the Calvin cycle,
    Which do not require light energy.

Energy to drive the Calvin cycle comes from adenosine triphosphate and reduced hydrogen carrier,
    Both generated in the light stage.

Carbon dioxide is picked up by 5-carbon ribulose biphosphate, RuBP (also known as ribulose diphosphate, RuDP).
RuBP reacts with carbon dioxide to form 3-carbon glycerate-3-phosphate, GP (also known as phosphoglyceric acid, PGA).
GP is converted to 3-carbon triose phosphate (TP) using ATP and reduced hydrogen acceptor.
About one-sixth of the formed 3-carbon triose phosphate is then converted into the 6-carbon sugar (hexose) glucose,
    Which may then be polymerised to form starch,
        To be stored or broken down later to yield energy,
    Or cellulose,
        To be used as a structural component of cell walls.
Remaining five-sixths of the formed 3-carbon triose phosphate is used to regenerate a supply of RuBP.
Six turns of the Calvin cycle are effectively needed to generate a molecule of glucose.
For every molecule of carbon dioxide fixed, three adenosine triphosphate and two hydrogen carrier molecules are used.

8 DISCUSS THE ROLE OF ATP IN LIVING MATTER

POINTS TO COVER

Energy is required by all living matter for biosynthesis, movement, etc.
Ultimate energy source is the sun,
    Which is used by green plants in photosynthesis,
        To form energy-rich chemicals, e.g. carbohydrates,
        Which may be used in respiration to release energy,
        Much of which is lost as heat,
        Which may speed up enzyme-catalysed reactions.

Energy released from respiration is not directly usable by cells.
Energy is made available in a controllable and usable form,
    Called adenosine triphosphate (ATP).
    ATP is a phosphorylated nucleotide compound.
ATP is known to be present universally in plants, animals, microbes, etc.
Almost all ATP synthesis occurs during oxidative phosphorylation phase of aerobic respiration.
ATP is capable of storing chemical energy released during respiration,
    Second phosphate bond in ATP is energy-rich (high energy bond) but it is the terminal linkage which is mainly responsible for energy storage.
On enzymatic hydrolysis of ATP to ADP and phosphate,
    Energy (about 30 kJ per mole) is made directly available to other reactions.
Not all the energy released during hydrolysis is used,
    Some is lost as heat,
    Remainder is then used in biosynthesis, nerve transmissions, muscular work, transport across membranes, etc.
Replenishment of store of ATP from ADP and phosphate,
    Requires further energy input from respiration.

Thus, ATP can be degraded or synthesised quickly where and when required in living cells.
ATP is not transportable.

9 **WRITE AN ESSAY ON MITOCHONDRIA UNDER THE FOLLOWING HEADINGS:**

(a) relationship between structure and function;

(b) occurrence of mitochondria in living cells.

POINTS TO COVER

(a) *Relationship between structure and function*

Sub-cellular organelles varying in shape from rod to spherical.
Surrounded by a double membrane,
    Inner membrane is granular,
        And highly folded to form cristae,
    Outer membrane is smooth.
Inner structure is the matrix,
    Where almost all of the enzymes of the tricarboxylic acid cycle are located.
Cristae have a large internal surface area,
    Which allows the attachment of many respiratory enzymes,
    Including those involved in the reactions of electron transport system,
        Whose enzymes are thought to be arranged in a sequence which reflects the order of the biochemical reactions,
        So that the metabolites do not have far to travel before engaging in the next stage.
Such reactions can occur 'outwith' cytoplasm,
    Main function is ATP synthesis.
Membranes are permeable to entry and exit of metabolites,
    Such as pyruvic acid, water, etc., involved in ATP synthesis.

(b) *Occurrence of mitochondria in living cells*

May be present in thousands per cell.
Cells which are metabolically active contain large numbers of mitochondria,
    For example, muscle cells, phloem companion cells, kidney tubule cells, sperm cells.
Generally, the greater the energy demand of a tissue, the more mitochondria its cells will contain,
    And the more extensively convoluted are their inner membranes.

10. IN WHAT WAYS DO THE AEROBIC AND ANAEROBIC PHASES OF RESPIRATION DIFFER IN ANIMAL TISSUE?

POINTS TO COVER

Aerobic Phase	Anaerobic Phase
Needs oxygen.	Does not need oxygen.
Occurs in mitochondria.	Occurs in cytoplasm.
Mostly cyclical pathway.	Mostly linear pathway.
Net gain of thirty-six ATP molecules.	Net gain of two ATP molecules.
Major mechanism for supplying energy to living cells.	Minor mechanism for supplying energy to living cells.
Cannot function when oxygen supply is insufficient such as during violent muscular activity; lactic acid not produced.	Can function for a short time during violent exercise to provide needed energy; lactic acid may be produced.
During exercise, rate of aerobic respiration decreases in skeletal muscle tissue.	During exercise, rate of anaerobic respiration increases.
Initial substrate is a 2-carbon molecule.	Initial substrate is a 6-carbon molecule.
Produces carbon dioxide.	Does not produce carbon dioxide.
Forms water.	Forms pyruvic acid.
Needs oxygen to produce ATP via cytochrome system.	Can produce ATP in absence of oxygen and cytochrome system.
Cytochrome oxidase necessary to force hydrogen and oxygen to combine to form water.	Cytochrome oxidase not necessary.

11 DISCUSS THE VARIETY OF PROTEINS AND THE ROLE THEY PLAY IN CELL STRUCTURE AND FUNCTION

POINTS TO COVER

Protein molecules are amongst the most important biochemicals,
    Found in almost every part of animal and plant cells.
Have very varied roles,
    But are commonly all made up of a combination of the same twenty amino acids.

Enzymes,
    Organic catalysts in living cells,
    Effect synthetic or breakdown reactions.

Structural components,
    Such as found in cell membrane,
    Or stroma of chloroplast.

Repair and replacement of worn out cell parts.

Can be metabolised as an energy source,
    In extreme starvation.

Growth.

Endocrine function,
    As hormones, such as insulin, which regulate cell function,
        Though not all hormones are proteins.

Buffers in the blood,
    Such as serum albumin, globulin, fibrinogen.

Movement,
    Such as contractile protein component of muscle cell filaments.

Nutrient store,
    As found in egg albumin.

Support,
> Such as collagen, keratin,
>> In skin cells,
> Or viral coat,
>> Which holds the nucleic acid core.

Pigment,
> Such as haemoglobin in red blood cells.

Carrier molecules,
> To transport materials into/out of cells.

Regulation,
> Controlling synthesis of other proteins,
>> Such as galactosidase,
>> By a protein repressor molecule.
> Or metabolism of glucose,
>> By insulin.

Defence,
> In antibody production.

12. WRITE AN ESSAY ON THE STRUCTURE OF THE DNA MOLECULE AND HOW THIS STRUCTURE ALLOWS REPLICATION TO TAKE PLACE

POINTS TO COVER

Nucleic acid found in the nucleus of plant and animal cells.
Very long biopolymer,
    Which is very stable,
    And capable of self-replication.

Basic structural unit is the nucleotide formed from,
    One of four nitrogenous bases,
        Adenine, guanine, cytosine and thymine,
        Which pair in a complementary fashion,
        Adenine with thymine,
        Guanine with cytosine,
    A five carbon sugar,
        Called deoxyribose,
    Phosphate group.

Sequence of nucleotides is the primary structure of DNA.
Sugar and phosphate form the backbone of the molecule,
    And are located on the outside.
The space between the backbones is that of two complementary nitrogenous bases,
    Which are located on the inside,
        And held together by weak hydrogen bonding.
Nucleotides are linked to form two chains,
    Which are complementary to each other.
Chains form a double helix,
    In which the two helices are wound round each other.
The helices are antiparallel,
    That is they have their terminal groupings at opposite ends,
    And their sequences run in opposite directions.
Linear arrangement of the bases in the structure,
    Is the basis of the genetic code,
    Having enough redundancy to code for all natural amino acids.

Replication of DNA starts with a breakage in the weak hydrogen bonding between complementary bases,
   Which exposes two naked single strands.
   Within the nucleus are the raw materials (nucleotides), energy supply (adenosine triphosphate) and enzymes (such as DNA polymerase which starts the replication process) needed to synthesise two new identical daughter strands.
   Each naked strand effectively acts as a template to allow complementation of the freely available nucleotides to begin building up new daughter DNA molecules,
   Which take up the same helical structure as the original parental DNA molecule.

13. WRITE AN ESSAY ON THE NUCLEIC ACIDS UNDER THE FOLLOWING HEADINGS

(a) the differences between DNA and RNA;

(b) the differences between messenger RNA and transfer RNA.

POINTS TO COVER

(a) The differences between DNA and RNA

Differences between DNA and RNA are significant in terms of their respective functions.

DNA contains the 5-carbon sugar deoxyribose,
    RNA contains the 5-carbon sugar ribose.

DNA contains the four nitrogenous bases, cytosine, guanine, thymine and adenine,
    RNA contains the base uracil instead of thymine.

DNA is a double stranded polymer,
    RNA is a single stranded polymer.

DNA exists usually as one type of molecule,
    RNA can exist as different types,
        Such as messenger, transfer, etc.

DNA is a very large molecule,
    RNA is a relatively smaller molecule.

(b) The differences between messenger RNA and transfer RNA

Although both molecules are transcribed from DNA, they have differing functions.

Messenger RNA (mRNA) carries genetic information,
    Which has been transcribed from DNA base sequence,
    To be translated into a protein,
    By ribosomes in cell cytoplasm.

Transfer RNA (tRNA) functions to transport specific amino acids into position for incorporation into a protein chain.

mRNA is highly variable in terms of size,
    Due to the diverse proteins for which mRNA codes.
tRNA is the smallest type of RNA,
    With much less variability in its structure.

mRNA accounts for less than 5% of the cell's total RNA,
    And has a relatively high molecular mass,
    With up to three thousand nucleotides.
tRNA accounts for about 15% of the cell's total RNA,
    And has a relatively low molecular mass,
    With an upper maximum of less than a hundred nucleotides.

mRNA is often found as a linear strand.
tRNA is usually highly folded on itself.

Three bases on mRNA form a codon,
    Which effectively codes for one specific amino acid.
Three bases on tRNA form an anticodon,
    Which distinguishes the different types of tRNA.

Breakdown of mRNA by enzymes in cytoplasm,
    Allows cell to control its metabolism.
tRNA is not susceptible to these enzymes,
    Because it has unusual bases not found in mRNA.

mRNA has a relatively short life span.
tRNA has a relatively long life span.

14. GIVE AN ACCOUNT OF THE STRUCTURE AND FUNCTION OF THE FOLLOWING SUB-CELLULAR COMPONENTS:

(a) ribosomes;

(b) Golgi apparatus;

(c) rough endoplasmic reticulum.

POINTS TO COVER

(a) Ribosomes

Function in protein synthesis.
Sub-cellular spherically shaped components,
    Found free in cytoplasm,
        Or attached to rough endoplasmic reticulum.
Composed of protein and nucleic acid,
    In roughly equal amounts.
Responsible for reading mRNA in cytoplasm (translation),
    One codon at a time,
    Each codon specifying one amino acid.
Ribosomes may be linked together to form a chain
    Called a polysome (or polyribosome).

(b) Golgi apparatus

Group of membranes lying alongside each other,
    Forming a number of flat sacs.
May be concerned with secretion,
    Such as secretion of cell wall material in young plant cells.
May be involved in formation of digestive enzyme containing vesicles,
    Called lysosomes.
Form carbohydrate-protein complexes,
    Such as mucin (which forms mucus).
Present in cells from almost every plant and animal.

(c) **Rough endoplasmic reticulum**

    ⎡ Three-dimensional network of membranes.
    ⎢ Often developed into slit-like cavities,
    ⎣   Called cisternae.

Membranes may function in controlling exchange of materials passing through the channels,
    Which may themselves form a transport system within the cell,
    As well as affording a large surface area for biological reactions.

Covered with ribosomes,
    Whose function is protein synthesis.

Found in large amounts in cells where demand for protein synthesis is high,
    Such as cells in enzyme-producing organs,
    Or actively growing cells.

15. WRITE AN ESSAY ON VIRAL INFECTION OF HUMAN CELLS UNDER THE FOLLOWING HEADINGS:

(a) natural invasion of human cells;

(b) natural barriers to viral invasion;

(c) alteration of host cells.

POINTS TO COVER

(a) Natural invasion of human cells

Viruses generally gain entrance to the body through mucus covered membranes of the breathing, digestive and urinary systems.
Others penetrate the skin,
    Or are injected by bites or cuts.

(b) Natural barriers to viral invasion

Body has a number of natural barriers to prevent entry via these routes.

Skin is a first line of defence,
    Having tightly packed epithelial cells,
    Acting as mechanical barrier,
    Preventing virus particles from gaining entry,
    While constant shedding of cells also removes any microbes.

Skin secretions
    Such as sebum secreted by sebaceous glands,
    Form a protective film over the skin,
    And contain chemicals such as fatty acids,
    Which inhibit microbial growth.

Mechanical flushing action,
    Of urine, tears, saliva,
    Helps protect membranes of urinary tract, eyes and mouth.

Many chemicals are antimicrobial such as,
    Bile and gastric juice found in the gut,
    Lysozyme found in tears.

Respiratory tract is lined with ciliated cells,
> Which, in concert with viscous mucus, trap inhaled virus particles,
> And sweep them up out of the lungs and associated air passages,
>> So-called 'ciliary escalator'.

Reflexes such as coughing and sneezing,
> Also help to forcibly expel virus particles which may have been inhaled.

During swallowing, the epiglottis closes the top of the trachea,
> Preventing microbes entering the lower part of the breathing system.

As air is drawn into respiratory tract, the air current which develops (vortex) tends to throw any particulate matter against the mucus/ciliated wall.

Nose is also lined with mucus covered hairs.

(c) *Alteration of host cells*

Viruses contain a nucleic acid core (DNA or RNA),
> Surrounded by a protein sheath,
> But no organelles.

In order to produce more virus particles,
> Viruses must utilise cellular machinery of a host cell,
> By injecting their nucleic acid,
>> Which instructs the host cell to assemble more viruses,
>> Which are then released to infect new host cells.

16 DISTINGUISH BETWEEN THE TWO DIFFERENT TYPES OF SPECIFIC IMMUNE RESPONSE BY HUMANS TO DISEASE

POINTS TO COVER

Specific immunity to disease is either active or passive.

Active immunity develops after exposure to, and recovery from, an antigenic invasion,
    The host producing specific antibodies to the antigen,
    In response to a natural encounter (naturally acquired active immunity),
    Or in response to deliberate exposure to antigen (artifically acquired active immunity),
        The basis of immunisation techniques,
    Normally long-lived immunity.

Passive immunity results from the introduction of antibodies into host,
    Which are made outwith the body of the host,
    Also may be natural,
        As is the case with maternal/foetal transfer of antibodies in colostrum,
        Or across the placenta,
    Or artificial,
        As happens when antibodies are (usually) injected into a host who is at immediate risk and may not be able to manufacture own antibodies quickly enough to survive exposure.
    Antibodies are normally produced in horse or another human,
    Normally short-lived immunity.

17. BY MEANS OF AN EXAMPLE YOU HAVE STUDIED, EXPLAIN HOW HUMANS CAN MOBILISE A WHOLE-BODY RESPONSE TO ANTIGENIC INVASION

POINTS TO COVER

Good example of such a response is the effect of receiving an organ transplant.

Normally, when a transplant is made, it will initially appear to be accepted,
> But within a few weeks, the transplant will be rejected,
> By the action of antibodies found on the surface of white blood cells,
> Which react with surface antigens on the transplanted tissue,
> Which eventually dies,
>> A second transplant from the same donor would be rejected within days.

This rejection process is immunologically based,
> Since the host recognises the transplant as foreign,
>> That is, antigenic,
> And initiates a specific immune response,
> Producing specific antibodies which destroy the transplanted tissue.

The measure of success of transplantation is directly related to the matching of the antigenic profiles of the donor and host tissues,
> This matching is normally at best only partial,
> Since each individual's profile is unique,
>> Except perhaps in the case of monozygotic twins,
> There is always going to be a rejection response,
> Requiring intervention, in the form of immunosuppressive drug therapy over a long period.

18 OUTLINE SOME OF THE MECHANISMS WHICH PLANTS HAVE TO AVOID INVASION BY FOREIGN BODIES

POINTS TO COVER

Unlike most animals, plants do not have the power of movement to avoid natural predators, seek more suitable environment, etc.
Not surprising therefore to find that plants have evolved a number of mechanisms to avoid invasion by foreign bodies.

Secondary plant substances,
    Which are chemicals found specifically in some plants but not others,
    Produced as by-products of normal plant metabolism,
        Include juglone (walnut trees), phenylpropanes (cinnamon, cloves, etc., in some herbs), terpenoids (e.g. peppermint oil), alkaloids (e.g. nicotine, morphine, caffeine),
    Which inhibit herbivore activity but have been shown to have pronounced antimicrobial activity,
        By inhibiting growth of pathogenic fungi,
        Such as potato blight, rust, etc.,
            Which are amongst the most common microbial agents of plant damage.

Mechanical protection,
    Includes the barriers of the cell wall and thickened waxy cuticle,
        Which normally are breached only by the action of biting insects,
        Or some kind of trauma, such as effect of grazers trampling or eating vegetation.
    Decomposition of lignin between cellulose of cell wall,
        Produces a tough protection which is difficult to invade.
    Secondary stem thickening to develop tough bark,
        Forms a relatively gas-tight covering which is a formidable barrier to foreign invasion.

19 DISCUSS HOW MITOSIS AND MEIOSIS DIFFER

POINTS TO COVER

Mitosis	*Meiosis*
One cell division.	Two cell divisions.
Nucleus divides once.	Nucleus divides twice.
Number and types of chromosomes in daughter cells same as in parent cell.	Number of chromosomes in daughter cells half that of parent cell and different in type.
Diploid number maintained.	Haploid number results.
Homologous chromosomes do not pair up.	Homologous chromosomes pair up (synapsis).
No chiasmata formed.	Chiasmata may occur.
No crossing-over, And no genetic exchange between chromatids.	Crossing-over, With possible genetic exchange between non-sister chromatids.
Two daughter cells formed.	Four daughter cells formed.
Occurs in body cells.	Occurs in gamete forming tissue.
No variation between daughter cells.	Variation possible between daughter cells.
Single line of chromosomes aligned on equator.	Double line of chromosomes aligned on equator.
Single spindle formed during whole process.	Two spindles formed during whole process.
One replication of centrioles if present.	Two replications of centrioles if present.

20. WRITE AN ACCOUNT OF SEX-LINKED INHERITANCE IN HUMANS

POINTS TO COVER

Chromosomes of males and females are not identical,
    In males the twenty-third pair is XY,
    While in females the twenty-third pair is XX,
    These are the sex chromosomes.
Sex chromosomes are not completely homologous.

As well as transmitting the primary sexual characteristic,
    Sex chromosomes also transmit other traits,
    Which may be controlled by genes found on the portion of the X chromosome absent in the Y chromosome,
    Producing an unusual inheritance pattern,
    Such as found in the inheritance of red/green colourblindness or haemophilia,
    Caused by recessive sex-linked genes.

Females must have two recessive genes to produce a sex-linked condition,
    While males can carry only one.
It follows that most sex-linked traits are more common in males than in females.

Other sex-linked traits include some forms of diabetes, a type of muscular dystrophy, some types of deafness, etc.

Sex-linked genes cannot be passed by a man to his sons,
    However a female carrier can pass disease-causing gene to her sons who would all develop the condition,
    While her daughters would be either normal or carriers,
        Depending on husband's genotype.

Well documented example of sex-linked inheritance is that of haemophilia in Queen Victoria.

21. DESCRIBE THE TWO MAIN FORMS OF MUTATION AND HOW THESE MIGHT ARISE

POINTS TO COVER

Two main forms are   1. Gene mutation;
  2. Chromosome mutation.

Natural forms of mutation are rare.
A mutation represents a sudden change in a gene or chromosome,
> Which may involve alteration in DNA molecule or an increase or decrease in the numbers of chromosomes.

Most mutations are harmful,
> Some are lethal.

Mutations may cause altered structure, physiology, biochemistry, fertility, lifespan, etc., of organism.
Usually having less than the normal number of chromosomes is more harmful than having more than the normal complement.

If mutation is in body cell, only individual is affected,
> However, if mutation is in reproductive organs,
> > All gametes produced may carry mutation,
> > > Thus potentially affecting many individuals or generations.

1.   Gene Mutation

   Not able to be detected microscopically.
   Involves change in linear arrangement of bases in DNA.
   May occur by loss (deletion), addition (insertion), error (substitution), rearrangement (inversion) of bases,
   > Such as occurs in sickle-cell anaemia,
   > > In which an altered protein component of haemoglobin is synthesised.

   Gene mutations may be induced by radiation, increasing temperature, certain chemicals such as mustard gas (called mutagens).

2. Chromosome mutation

May be detectable microscopically.
Involves change in the normal number of chromosomes.
May occur as a result of loss (deletion), rearrangement (inversion), relocation (translocation), self-copying (duplication) of chromosome fragments,
> Or failure of chromosomes to separate during meiosis (non-disjunction),
> As in the case of polyploidy,
>> Where several sets of chromosomes instead of normal number are present,
> Or mongolism,
>> Where there is an increase of one chromosome (number 21),
> Or Klinefelter's syndrome,
>> Where there is an extra sex chromosome (XXY),
> Or Turner's syndrome,
>> Where a sex chromosome is missing (XO).

May be induced by use of drug colchicine,
> Which prevents spindle formation.

22 *A KNOWLEDGE OF GENETICS CONTRIBUTES TO IMPROVEMENTS IN AGRICULTURE. BY MEANS OF EXAMPLES YOU HAVE ENCOUNTERED, DISCUSS THIS STATEMENT UNDER THE FOLLOWING HEADINGS:*

(a) plant breeding;

(b) hybrid vigour in animals.

POINTS TO COVER

Without a clear understanding of genetics, agricultural developments would be very slow.
Such understanding allows humans to obtain desired results by selective breeding much quicker than by natural selection.
Individuals not meeting the specifications are prevented from breeding.

(a) Plant-breeding

    Humans can artificially select plants,
        From wild strains such as *Brassica oleracea*,
        To give new 'artificial' varieties,
        Which have a particular desired phenotype,
        Such disease resistance, insect resistance,
        Or an ability to tolerate extremes of weather and climate,
        Or give high yield, etc.
Poorer countries of the world may now grow crops which are high-yielding and disease resistant.
Plant-breeding forms part of the 'green revolution',
        That is, the advances in agricultural practices.

(b) Hybrid vigour in animals

    Outbreeding between unlike animals may produce hybrid vigour,
        So that the hybrid is better than either parent,
        Which may increase meat or milk yield, quality or quantity of wool, etc.

For example, mules are the result of crossing a donkey and a horse,
> While horses are faster and stronger than donkeys,
> The latter has the greater stamina.

A mule combines the desirable features of stamina and speed,
> But is incapable of interbreeding to produce more mules.

Another example is the selection and interbreeding of dogs,
> Such as labrador and retriever for use as guide dogs by the blind,
> Combining qualities of the two varieties,
> And increasing yield of suitable pups to upwards of 80%,
> Or alsations as 'sniffers' by Customs to detect drugs/explosives.

23. WRITE AN ESSAY ON EVOLUTION UNDER THE FOLLOWING HEADINGS:

(a) adaptive radiation;

(b) natural selection.

POINTS TO COVER

(a) Adaptive radiation

Since food, available space, etc., are limited,
>Organisms have evolved to occupy almost every possible habitat,
>A process called adaptative radiation.

Adaptative radiation is advantageous,
>Since it allows organisms to exploit new resources in the environment.

One of the best known examples is that of 'Darwin's finches',
>Found on the Galapagos Islands.

These finches were derived from a single ancestral mainland stock,
>But now show huge variations in their beak size and function and feeding habits,
>Some feeding on insects, others on seeds, cacti, etc.,
>Allowing each species to occupy a unique ecological niche,
>And exploit a particular food source.

(b) Natural selection

Mechanism put forward by Charles Darwin (1809–1882),
>To explain evolution by natural selection.

Based on number of observations and deductions,
>Most organisms over-produce (angiosperm seeds, fish eggs, etc.),
>Yet population numbers are relatively constant,
>Thus many offspring do not survive to reproduce,
>Competition will exist amongst offspring for space, food, light, etc.,

Variation exists in living things,
Those individuals most or least suited to environment have best or least chance of surviving to reproduce,
If variations are inheritable, offspring will also have the favourable genotype with its resultant favourable phenotype,
    If this natural selection process is repeated it will result in accumulation of such advantages in organisms belonging to the same species,
    And may eventually result in the formation of a new species.

24 *DISCUSS EVOLUTION TODAY BY REFERRING TO EACH OF THE FOLLOWING:*

(a) industrial melanism;

(b) resistance.

POINTS TO COVER

(a) Industrial melanism

Allows natural selection to be observed in action,
    Studied extensively by Kettlewell in England in the 1950s.

Moth, *Biston betularia*, can exist in two forms (polymorphism),
    The light-coloured variety found to be rarer than dark-coloured variety in industrialised areas.
Moth is normally found on tree trunks during day.
Before industrial revolution trunks were normally lichen-covered and light-coloured,
Normal light-coloured moths were not noticed by predators (insectivorous birds),
    And their numbers were high,
Dark mutant forms were easily seen,
    And their numbers were low,
With industrialisation trunks became soot-covered and dark,
Making the normal light-form noticeable to predators,
    And so their numbers fell.
Dark forms now were difficult for predators to see,
    And so their numbers started to increase,
This process is now reversed with increasing attention to prevention of air pollution.

Similar events have been observed in other animals when background colour of environment has changed.

(b) Resistance

Resistance may develop to antibiotics, insecticides, toxic chemicals, etc.,
> Involves very rapid selection of resistant forms,
> By selective agent,
>> Such as penicillin, DDT, malathion, etc.,
> Non-resistant forms are killed (selected against) by agent,
> Leaving resistant forms to survive and reproduce.

Individuals (bacteria, houseflies, rats, etc.) already possess the mutation giving resistance,
> Resistance is not caused by agent,
>> This can be shown by producing pure culture of antibiotic resistant bacteria which have never been in contact with antibiotic.

May pose serious health hazards.
Danger occurs if an agent is over-used,
> Or used when situation is not life-threatening or serious,
>> Possibly resulting in the selection of bacteria having multiple resistance.

25 HOW MAY A SPECIES BECOME EXTINCT?

POINTS TO COVER

Extinction is a natural phenomenon,
   Which has occurred ever since life began.
   Some estimates suggest that 98% of all life that ever existed has now become extinct,
   As supported by fossil evidence,
      Such as skeletal remains, footprints, preserved specimens (amber, ice, etc.).

So-called 'mass extinctions' which destroyed up to 90% of all living species are now well known,
   Such as 'sudden' disappearance of dinosaurs,
   Which preceded the age of mammals.

Competition for food and resources, etc.,
   Means that two different species cannot occupy the same niche,
   Which might cause extinction of one.

Habitats can be drastically altered,
   By sudden global climatic changes,
   Leading to extinction of species which cannot adapt.

Introduction of a new predator, disease, etc., may completely eliminate a species.
Recent history records the extinction of several species,
   Such as the dodo, great auk, etc.

Mass killings for food and sport have brought some species close to extinction,
   Such as the American bison, white rhino.

The exploitation of particular environments,
   Such as tropical rainforests,
   Will be destroying species which may not even be catalogued.

Indiscriminate use of toxic chemicals or improper elimination of industrial wastes may be destroying aquatic species.

26. BY MEANS OF EXAMPLES, SHOW HOW THE ABILITY TO MANIPULATE GENES CAN BE USEFUL TO HUMANS

POINTS TO COVER

Gene manipulation depends on a knowledge of how to alter the nucleic acids, DNA and RNA, of organisms,
    By inserting, deleting or modifying one or more genes,
    In the cell(s) of a particular organism,
    Giving rise to a new phenotype which will have some desirable property.
This is the basis of genetic engineering and biotechnology.

May be exploited by humans in a number of ways:

1. Production of a desired chemical,
    Such as the hormone insulin or human growth hormone, enzymes, drugs (interferon), vaccines, blood clotting proteins (factor VIII), antibodies, etc.,
    By a genetically engineered organism, such as the bacterium *Escherichia coli*.

2. Production of desirable phenotypes in plants,
    Giving rise to disease resistance, particular taste-qualities, hybridisation, colours, etc.

3. Possible alleviation from diseases which have a genetic basis,
    Not amenable to conventional therapy,
    Such as sickle-cell anaemia, cystic fibrosis, etc.,
    Where the possibility now exists, because of human gene mapping techniques,
    Of detecting and altering the affected gene or replacing it.

4. Enhancing food production,
    In developing new high-yielding strains of crops, cattle, sheep, etc.,
    Or introducing genes which give crops the ability to fix atmospheric nitrogen,
    Or producing food from commercial culturing of genetically engineered micro-organisms (fungi, bacteria, algae, etc.) to give single-cell protein.

27 DESCRIBE HOW A NEW SPECIES MIGHT ARISE

POINTS TO COVER

In any species, individuals may be able to breed with other members of the same species.
Members of the same species resemble each other very closely (morphologically, physiologically, etc.),
    Forming a population of related organisms,
    Which cannot breed with members of a different species to produce fertile offspring.
Offspring produced by successful breeding between members of the same species are fertile,
    Such offspring may freely interbreed.

Speciation involves splitting of the group constituting the species.
Splitting must be followed by isolation of parts of the groups,
    Which ensures exchange of gametes cannot occur,
    And genetic differences are maintained,
    While mutations and selection can operate separately and differently on the isolated sub-groups.

Isolation experienced by the sub-groups may be
    Ecological (habitat, seasonal differences, etc.),
    Reproductive (no interbreeding possible, behavioural differences, etc.),
    Geographic (mountains, rivers, etc.),
    Genetic (offspring sterility, etc.).

If new group formed cannot breed successfully with original,
    To produce fertile offspring,
    Then speciation has occurred.

Process can be reversed if barriers are removed.

28 DISCUSS THE ROLE OF MERISTEMS IN HIGHER PLANTS

POINTS TO COVER

Meristematic tissue is found in main, or lateral, root and shoot tips (apical meristem) and in cambium (lateral or secondary meristem).

Meristematic cells are relatively thin walled,
    With large nucleus,
    And reduced vacuole or no vacuole at all.
Meristematic cells are responsible for formation of new cells by mitotic divisions,
    And are capable of producing all mature plant tissues.
Meristematic tissue becomes active very early in life of plant,
    Can re-grow if damaged,
    Permits growth throughout life of higher plants,
    And is a major difference between plant and animal growth.

Increase in thickness (girth) of stem or root due to cambial activity (secondary growth),
- Cambium is a lateral meristem between xylem and phloem of vascular bundles,
- Which eventually meets to form a cylinder,
- And can divide to produce new cells,
- Differentiation occurs into secondary xylem (inside) and secondary phloem (outside),
- Continued cambial activity forces phloem (and associated cambium) to edges and girth increases (secondary thickening).

Original epidermis is replaced by cork cambium arranged circularly.

Meristems are subject to the effects of weather,
    So that increasing temperature, for example, will result in increased activity of meristems.

Meristems form leaf embryonic leaves (primordia).
Shoot and root apical meristems are protected by apical bud and root cap respectively.

Increase in length of root and shoot is caused by activity of apical meristem,
    Such increase is called primary growth.
It is thought possible that apical meristems may influence the development of other tissues.

29 CONTRAST GROWTH AND REGENERATION IN ANGIOSPERMS AND MAMMALS

POINTS TO COVER

In angiosperms, growth and cell differentiation overlap,
    So that, for example, apical growth may precede cell differentiation,
    Whereas in mammals growth takes place after cell differentiation is finished.

Growth in angiosperms is continuous throughout the plant's life,
    Whereas in mammals growth stops at adulthood (shrinkage may occur in old age).

In mature angiosperms, growth occurs in localised areas,
    Whereas in mammals growth is diffused.

In angiosperms, growth involves specific tissues called meristems,
    Whereas in mammals all tissues and organs may be involved.

In angiosperms, growth results in a larger surface area and branched appearance,
    Whereas in mammals growth results in relatively compact body (except for legs and arms).

Regeneration is the growth of organs, tissues or cells which have been damaged or removed.
In angiosperms, regeneration is extensive,
    For example in the activation of dormant tissue or secondary meristems,
    So that a whole plant can be regenerated from a fragment of the original,
    Whereas in mammals such regeneration is highly restricted,
    For example to wound healing, nerve fibre growth after damage, liver, thyroid and pancreatic growth after partial removal.

In angiosperms, regeneration is not dependent upon a nervous supply,
    Whereas in mammals regeneration is thought to depend heavily on the nervous supply to the damaged area.

30. BY REFERRING TO EACH OF THE FOLLOWING, EXPLAIN HOW THE GENOTYPE CAN INFLUENCE GROWTH AND DEVELOPMENT

(a) development of cell;

(b) development of an organism.

POINTS TO COVER

(a) Development of cell

How a cell develops is a direct function of its genotype,
Excepting gametes, all the somatic (body) cells of a multicellular organism are normally genetically identical,
But the expression of its genotype will vary from cell to cell during growth and development,
By the switching on or off of particular genes,
Which may depend on other genes being present (genetic environment),
And/or the physical environment within which the gene expression takes place (temperature, humidity, light, etc.),
Cells inherit potentiality whose expression is a product of genotype and environment,
At the individual cell level, the expression of genes may vary according to other factors.

Examples include the development of human red blood cells (erythropoiesis) from primitive stem cell in red bone marrow,
In which the genes controlling the synthesis of haemoglobin and carbonic anhydrase (involved in carbon dioxide transport) are switched on while other genes are switched off,
And the development of parenchyma cells from relatively undifferentiated meristematic cells,
Which can become modified to become photosynthetic by the switching on of genes controlling development of chloroplasts,
To become mesophyll.

(b) ***Development of an organism***

Since mitosis gives rise to daughter cells whose genetic complement is the same,
> All the body cells in a multicellular organism are genetically identical.

There is no evidence to suggest that during development of an organism any genes are lost,
> So that the particular development of embryonic tissues and organs must be reflection of combined effects of environment and collective gene expression of groups of cells.

Seen in the development of human skin colour,
> Due to the presence of melanin,
> Secreted by cells called melanocytes,
> Which are found in roughly the same numbers in individuals from all races,
> Suggesting that in some races, there are differences in genotypes which affect melanin synthesis,
> Which in fact is an example of polygenic inheritance,
> That is, where two or more independent genes have similar and additive effects on the same phenotypic trait.

31 HOW IS HUMAN GROWTH AND DEVELOPMENT AFFECTED BY THE PITUITARY GLAND?

POINTS TO COVER

Pituitary gland influences growth and development via two hormones,
    Growth hormone and thyroid stimulating hormone.

Growth hormone (GH),
    Affects general body growth,
    Since it activates growth in body cells.
Principal effects of growth hormone are on the skeleton,
    And on the muscles of the skeleton,
    Causing an increase in their growth rate,
    And maintaining their size in adulthood.

Growth hormone is thought to stimulate cells to take up amino acids,
    And increase protein synthesis.
If insufficient growth hormone is produced in early life, bone growth is impaired,
    Causing dwarfism to develop,
    While other organs may fail to develop properly.
If too much growth hormone is produced in early life, bone growth is excessive,
    Causes gigantism to develop.
If too much growth hormone is produced in adulthood, agromegaly results,
    Causing thickening of bones of hands, cheeks, jaws, etc.

Thyroid gland growth and endocrine function are both controlled by thyroid stimulating hormone (TSH),
    Secreted by the pituitary gland.
Thyroid stimulating hormone induces the thyroid to secrete the hormone thyroxine,
    Depending on blood levels of thyroxine, glucose and the body's rate of metabolism,
    Which regulates growth, development and differentiation of cells,
    And works with growth hormone to stimulate body growth.

32. DISCUSS THE INFLUENCE OF INDOLEACETIC ACID ON PLANT GROWTH WITH REFERENCE TO THE FOLLOWING:

(a) effect on cells and organs;

(b) practical applications.

POINTS TO COVER

(a) Effect on cells and organs

Indoleacetic acid belongs to the group of plant hormones collectively called auxins,
> And has profound effects on plant cells and organs,
> Such as phototropic response of shoots,
> Caused by uneven distribution of the hormone in the stem below the apical meristem producing it,
> The cells receiving the higher concentration of IAA becoming more elongated, producing bending response towards light.

Auxins also are responsible for inhibiting the growth of lateral buds in plants,
> A phenomenon known as apical dominance,
> Caused by auxin production in the terminal bud,
> Travelling downward at a concentration which both stimulates stem elongation,
> While inhibiting lateral bud growth.

Auxins may also cause fruit to develop,
> By promoting the continuing growth and development of the ovary and receptacle after fertilisation.

Another property of auxins is to bring about shedding of leaves, flowers and fruits from the parent plant,
> When the auxin concentration at the bases of these structures falls below a critical supportive threshold,
> And abscission then occurs.

(b) Practical applications

Auxins are of tremendous commercial importance because their practical applications are so diverse:

Promotion of root growth,
    In cuttings.

Production of parthenocarpic fruits,
    That is, fruits produced without pollination,
    By spraying the female flowers,
    To produce seedless oranges and grapes.

Promotion of fruit growth,
    Since fruits grow larger and quicker if the parent plant is auxin treated.

Action as weed-killer,
    Particularly dicotyledonous plants,
    Which are stimulated to accelerate their growth until they exhaust their cellular reserves and die.

33 BY MEANS OF EXAMPLES, SHOW HOW THE ENVIRONMENT CAN INFLUENCE GROWTH AND DEVELOPMENT

POINTS TO COVER

Environmental factors may prevent/allow achievement of genetic potential.

Nutrition, dietary deficiencies may result in poor child growth,
    Even if that child has the genetic potential to be tall and well-developed,
    For example, poor vitamin D intake will result in skeletal malformations.

Drugs, if able to pass across human placenta into developing foetus,
    May interfere with the proper development (teratogenesis),
    Includes thalidomide, some antibiotics, antitumour agents, etc.

Alcohol, suspected teratogen for long time,
    Now associated with foetal alcohol syndrome (FAS),
    Which causes slow growth before and after birth, small head, defective heart, etc.,
    If mother abuses alcohol during pregnancy.

Nicotine, all evidence supports a relationship between cigarette smoking during pregnancy and poor foetal growth,
    As well as foetal mortality.

Light, direction of light affects root and shoot growth,
    Lack of light causes etiolation,
    And may prevent chlorophyll synthesis,
    Also involved in flowering response,
    While in many mammals lack of light may prevent vitamin D synthesis,
    And photoperiod may be involved in seasonal responses,
    Such as development of sexual organs during breeding season.

Gravity, affects direction of shoot and root growth.

Temperature, plant growth may be accelerated at high temperatures,
    While low temperatures may break dormancy.

Hormones, deficiency or overproduction of pituitary growth hormone can cause dwarfism or gigantism respectively,
    While in plants can induce/inhibit growth.

34. GIVE AN ACCOUNT OF MINERAL NUTRITION IN PLANTS UNDER THE FOLLOWING HEADINGS:

(a) function of macroelements;

(b) experimental evidence for the importance of macroelements.

POINTS TO COVER

(a) Function of macroelements

Macroelements are nutrients which are required in relatively large amounts for healthy growth of plants,
    Include nitrogen, potassium, phosphorus and magnesium.
Nitrogen is an essential element for synthesis of proteins, nucleic acids, adenosine triphosphate, chlorophyll, etc.
Potassium contributes to proper three-dimensional shape of proteins,
    And affects some enzyme-catalysed synthetic reactions.
Phosphorus is an essential element in the formation of the sugar-phosphate backbone of nucleic acids, adenosine triphosphate, cell membrane phospholipids, etc.
Magnesium is essential for chlorophyll function.

(b) Experimental evidence for the importance of macroelements

Relative importance can be demonstrated by growing plants in different water cultures,
    Whose chemical content is known exactly and can be varied.
Culture media are made each deficient in one macroelement.
Good subjects are seedlings (such as grasses),
    Which have small reserves of food,
        And therefore contain only small amounts of stored macroelements.
Gaseous exchange must be possible between culture medium and atmosphere,
    Which can be effected by aeration of the culture solution.
A water-tight cork prevents evaporative loss of medium.
All seedlings must have reached a similar stage of growth.

Light-proof paper covers each tube,
> Thus preventing the growth of algae.

Growth is allowed to proceed for several weeks.

A control culture should be used,
> Having all the macroelements present,
> In the correct proportions.

Distilled water is used which has no mineral content,
> To eliminate possible source for growing plants.

Visual differences such as chlorosis and stunted growth will appear,
> Which can be followed by comparison of dry-weights.

All cultures must be subjected to the same environmental conditions of temperature, etc.

35 HOW IS FLOWERING IN PLANTS AFFECTED BY THE DAYLENGTH?

POINTS TO COVER

It has been known for over sixty years that the length of day can affect flowering plants.
The specific length of day required is called the photoperiod.

Some plants are short-day plants,
    Which flower only when length of daylight in 24 hours is below a critical value,
    Included poinsettia, chrysanthemum, rice, etc.,
    Which tend to flower in winter.

Other plants are long-day plants,
    Which flower only when length of daylight in 24 hours is greater than a critical value,
    Includes most tobacco plants, barley, wheat, lettuce, etc.,
    Which tend to flower in summer.

A few plants do not need specific day lengths,
    And are the day-neutral plants,
    Which are unaffected by length of daylight in 24 hours,
    Includes some tobacco plants, tomato, cucumber, etc.

Good evidence suggests that it is the length of the dark period which is critical for flowering,
    Especially in short-day plants.
Blue pigment phytochrome is active in these responses.
Phytochrome can exist in two different forms.
When exposed to light of wavelength 660nm (red),
    It changes into a light-green form which absorbs only far red light,
        Which form is termed P735.
When phytochrome is in darkness,
    It slowly reverts back to red-absorbing pigment,
        Which form is termed P660.

In a short-day plant when the length of darkness is sufficient for P735 to be converted to P660,
    The plant will begin to flower,
    The length of darkness being needed about twelve hours.
If however darkness is interrupted by a flash of light,
    P660 is immediately converted to P735,
    And plant will not flower.

Long-day plants need P735 which is formed normally during long days and short nights.
Suggestion is that a hormone is involved in the flowering response,
    Called florigen,
    Which may be the initiator of flowering.

Defoliation of a plant leaving only one leaf still allows normal photoperiodic response,
    But not if defoliation is complete,
    Which supports the theory that the leaf itself is essential for flowering response.
After production in leaf, florigen may be transported in phloem,
    To affect buds.

36 DISCUSS THE ROLE OF THE SKIN AND HYPOTHALAMUS IN MAINTAINING A CONSTANT BODY TEMPERATURE

POINTS TO COVER

Skin is very important in body temperature maintenance,
- Nearly 80% of all heat loss occurring through the skin.

Skin itself is a poor conductor of heat,
- And thus helps prevent heat loss,
- Especially if the subcutaneous fat layer is thick.

Normal body temperature is about 37 °C.

Temperature-sensitive receptors in hypothalamus monitor blood temperature,
- And produce appropriate physiological responses to compensate,
- Acting as a thermostat.

Information is picked up by temperature-sensitive receptors in the skin,
- Which is then conducted via sensory fibres to the hypothalamus.

If temperature rises above normal, skin capillaries dilate,
- And more blood passes near the surface,
- To increase radiative heat loss,
- Tiny hair muscles relax,
  - So that hairs lie flat on skin,
  - Which prevents warm air forming a layer over skin.

Present in dermis of skin are many sweat glands.

If temperature rises above normal, sweat production is increased,
- And evaporation of sweat causes local cooling.

If temperature falls below normal, skin capillaries constrict,
- And less blood passes near the surface,
- Decreasing radiative heat loss,
- Tiny hair muscles contract,
  - So that hairs are erect on skin,
  - Which traps an insulating layer of air over skin,
- Sweat production also decreases,
- While involuntary shivering occurs,
  - Resulting in extra heat production.

37 BY MEANS OF ONE EXAMPLE, EXPLAIN WHAT IS MEANT BY A HOMEOSTATIC MECHANISM

POINTS TO COVER

A homeostatic mechanism allows animals to preserve a constant internal environment,
    With respect to such variables as temperature, blood sugar levels, hormone output, etc.
A constant internal environment provides optimum conditions,
    For particular processes,
    Or body parts to work.
This internal equilibrium state is dynamic and is not fixed,
    Requiring constant monitoring,
    Of changing conditions,
    By specialised receptors,
    And feedback,
    To particular effectors,
    To restore the equilibrium state.

Homeostatic mechanisms are highly developed in mammals.
Good example of homeostatic mechanism is osmo-regulation by the kidneys.
Tissue fluid is normally about 98% water.
If water content rises or falls much above or below this figure the body cells will respectively,
    Swell up with water,
    Or shrink with loss of water.
Osmotic pressure results from water and dissolved substances.

Changes in the osmotic pressure of the blood,
    Are detected by receptors in the brain.
When the blood osmotic pressure rises above normal,
    The pituitary gland is stimulated to produce anti-diuretic hormone (ADH),
    Which is transported in the blood plasma.

ADH acts on the kidneys,
>	Causing them to reabsorb more water,
>	Which increases urine concentration,
>	And decreases blood osmotic pressure.
> If the blood osmotic pressure falls below normal,
>	ADH production is inhibited,
>	And less water is reabsorbed,
>	Which produces very dilute urine,
>	And increases blood osmotic pressure.

The operation of this homeostatic mechanism is closely influenced by daily body activity,
>	Such as exercise, fluid intake, etc.

38. OVER A LONG PERIOD OF TIME, A POPULATION SIZE USUALLY REMAINS STABLE. EXPLAIN WHY THIS HAPPENS

POINTS TO COVER

A stable population usually consists of a large number of freely interbreeding individuals,
    Where mating is completely random.
In any stable population, mature individuals are capable of reproducing offspring,
    At a rate which is potentially exponential,
    If the environment is suitable,
        That is provides unlimited space and food, etc.
However environment is not able to support such exponential growth,
    Because there will be competition for resources which are not limitless,
        Such as food, shelter, breeding places,
        And organisms may be unable to adapt or search out alternatives,
    Rate of predation may increase,
    Disease may curtail numbers.
Eventually a limiting size is reached,
    Called the carrying capacity of the environment,
    Where death rate and birth rate are in balance,
    And the population size remains stable.

Homeostatic behavioural changes,
    Called epideictic behaviour,
    May occur in a population if size begins to move over the carrying capacity,
    Causing a reduction in birth rate,
    Or an increase in dispersal behaviour.

Small fluctuations in population size may occur,
    Which might cause a temporary increase or decrease in the short term,
    But the effect of these will be negligible in the long term.

In a stable population, it has been found that common species remain common while rare species remain rare.

39. WRITE AN ESSAY ON THE FACTORS WHICH INFLUENCE POPULATION CHANGE UNDER THE FOLLOWING HEADINGS:

(a) density dependent factors;

(b) density independent factors.

POINTS TO COVER

(a) **Density dependent factors**

The density of a population is the number of individuals in a given area.
As the density increases, factors come into operation which tend to control this change,
    And maintain the population size at the carrying capacity of the environment.
The density of predators is related to the density of the prey,
    As latter increases/decreases so also does the predator density,
    Tending therefore to maintain the population size.

Individuals may move out of the environment (emigration effect),
    Reducing the density of an increasing population,
    Which is a behavioural response that is density dependent.

Competition will become more evident as the numbers increase,
    For limited resources.

Physiological changes may also occur as population size increases,
    So that normal body defence mechanisms (immune response, etc.) are diminished,
    Leading to an increased susceptibility to disease.

(b) Density independent factors

Possible for controlling factors to exert effect long before population density increases above carrying capacity.
These are independent of density,
    And mainly include physical or abiotic factors.
May take the form of a sudden change in the environment,
    Such as severe storm, fire, flooding, volcanic explosion,
    Which are wholly independent in their action of the density of the affected organisms,
    The average percentage destruction remaining the same no matter how dense/sparse the population.

Density independent factors are thought to be important in controlling the size of organisms whose life-cycles are short.

40 DISCUSS SUCCESSION AND CLIMAX IN PLANT COMMUNITIES

POINTS TO COVER

Plant communities are rarely fixed,
    But more usually constantly changing,
    Becoming more complex,
    As one species replaces another, etc.
Succession is this orderly change in time of the plant communities in a particular area,
    So that for example a lake may fill up and eventually become a forest,
    Or newly exposed bare ground becomes colonised by a series of different plant species,
    Each stage being called a sere.
Resultant effect is an ecosystem which is relatively stable and balanced,
    Called the climax community.
Thus succession exhibits a direction and is predictable.

First stage in succession is colonisation by simple species,
    Called pioneers,
    Such as grasses, lichens, mosses, etc.,
    Which experience little or no competition,
    And whose action changes the environment, allowing more advanced species to colonise.
Slowly the pioneers are replaced by more advanced species,
    But it is believed that change is more rapid in the early stages than in the later stages.
Process continues until the climax community becomes formed,
    Which is usually quite complex,
    With large broad-leaved trees,
    And will remain stable,
        Perhaps lasting for hundreds of years,
    Unless some external factors operate,
        Such as climatic changes, grazing, etc.

Primary succession involves colonising newly exposed bare ground.
Secondary succession involves an area being colonised which was already inhabited,
    But no longer is because of the action of agent(s) such as fire, storms, etc.,
    Accounting for most of the present day climax plant communities in this country.

41. DISCUSS POPULATION CHANGE UNDER THE FOLLOWING HEADINGS:

(a) *population monitoring;*

(b) *factors governing population size.*

POINTS TO COVER

(a) Population monitoring

Population scientists monitor animal and plant populations for several reasons,
> Organisms concerned may be a valuable food resource,
>> Such as fish, deer, etc.,
> Or a potential threat to existing food resources,
>> If numbers exceed some critical threshold,
>>> Causing damage or loss.

To take conservative action if there is a predicted danger of numbers falling,
> So that extinction is a possibility,
> As in whale populations, protected plant species, etc.,
>> Which is happening in loss of tropical rainforests,
> With a permanent reduction in genetic diversity.

Usefulness as indicator species,
> In demonstrating the quality of water,
> By absence or presence of particular invertebrates,
> Such as bloodworms, mayfly nymphs, etc.
> Levels of atmospheric pollution,
> Such as lichen which are particularly sensitive to sulphur dioxide levels.
> Or the effect of some environmental change,
> Such as accidental spillage of chemicals, oil, etc.

(b) Factors governing population size

Anything which increases or decreases an organism's chances of surviving and reproducing will affect population size. Parasites and disease can exert heavy pressure on a population,
> Such as malarial parasite on human populations in parts of Africa, myxomatosis virus on rabbit populations in Europe, excessive defoliation of plants by herbivores.

Predators normally do not have a long-term effect on population size,
    Unless they are particularly efficient,
    Or for some reason the prey population is slow to recover.
Availability of nutrients is thought to be the single most critical factor governing population size,
    Plants require sufficient light, carbon dioxide, water and mineral nutrients to photosynthesise and grow,
        Drought conditions for example could seriously affect plant populations.
    Animals require abundant nutrients to have good chance of survival and reproduction,
    If nutrients are scarce, much time is needed to find adequate nutrition,
    If quality of food is poor there may be insufficient material/time to obtain enough nutrients.
Other variables which can effect population change include,
    Weather and climate,
    Territorial availability,
    Dramatic environmental changes (forest fires, earthquakes, etc.).

42 WHY IS WATER SO IMPORTANT IN THE BIOLOGICAL WORLD?

POINTS TO COVER

Evidence suggests that life evolved in water,
    And so it is logical to expect close links between living things and water.
Many living things spend their entire lives in water.
Most organisms have a very large body water content,
    Which can be up to 99% in the case of jellyfish.
Body metabolism is closely linked to water.
Water is the universal solvent,
    Which makes it important in facilitating biochemical reactions to take place.
Water may be a reactant or product in biochemical reactions,
    Such as photosynthesis or aerobic respiration.
It is a major source of hydrogen and oxygen,
    To be incorporated into other molecules.
It is a medium for transporting materials,
    Within the bodies of animals and plants.
Water has a high specific gravity,
    Which means that organisms will float easily in water,
    Thus water can act as a support for large aquatic life forms.
Reproduction in many animals and some plants depends on water,
    As a medium for carrying gametes.
Since pure water is transparent,
    Light can penetrate it for photosynthetic algae.
A large body of water is very heat stable,
    So that temperature changes in many aquatic environments are minimal.
Water has its greatest density at 4 °C,
    Which means that ice will float,
    And organisms living near the bottom are unaffected.
Osmotic effects depend on movement of water across semi-permeable membranes.
Uptake of mineral salts by roots and plants,
    From soil occurs in aqueous solution.
Temperature regulation is often water based,
    Such as sweating response.
Elimination and dilution of toxic wastes often occurs in solution.
Many natural lubricants are water-based.

43 HOW IS WATER-BALANCE MAINTAINED IN A:

(a) marine bony fish;

(b) desert mammal?

POINTS TO COVER

Regulation of body water content (osmoregulation),
    Is connected with control of osmotic pressure of body fluids.
This osmoregulation is often combined with the removal of nitrogenous wastes.
Structures responsible for osmoregulation are often the organs of excretion,
    Such as the kidneys.
Osmoregulation is an energy-demanding physiological process.

(a) Marine bony fish

Typical example is cod,
    Which has body fluids which are hypotonic to sea-water,
    Such fish lose water constantly by osmosis,
        Mainly across the gills,
    The volume of filtrate from kidneys being very small,
        And very concentrated as a result.
    To compensate for fluid loss, marine fish swallow sea-water,
    And expel excess salt,
        By means of chloride secretory cells in the gills,
    This salt being moved against a concentration gradient,
    And therefore requiring energy to operate active transport mechanism.

(b) Desert mammal

Typical example is the kangaroo rat,
- Whose water gain or loss is so well-regulated that animal rarely drinks,
- Water is lost through breathing, defecation and urination,
  - But is balanced by metabolic water and water in food.
- Regulation is by means of the kidneys,
  - Which are highly efficient and produce very concentrated urine,
  - Via long loops of Henlé,
  - Which absorb a great deal of water.
- Hormonal control of kidney function by means of antidiuretic hormone,
  - Which is usually present in high concentrations.
- Osmoreceptors in the brain detect changes in blood osmotic pressure.
- Behavioural adaptations,
  - Include avoidance of intense heat during day.
- Physiological adaptations,
  - Include long loops of Henlé and long collecting ducts.

44. GIVE AN ACCOUNT OF HOW WATER IS TAKEN UP BY A MATURE OAK TREE AND ITS PASSAGE THROUGH THE STEM UNTIL IT IS LOST AS WATER VAPOUR TO THE ATMOSPHERE

POINTS TO COVER

Unless the atmosphere is 100% saturated,
    An oak tree will lose water by evaporation,
        Mainly through its leaves.
This process is called transpiration,
    And is thought to be the main factor in the transportation of water in land plants.

Transpiration through leaves assists in drawing water upwards.
Osmotic gradient across the mesophyll cells draws water from the xylem.
Spongy mesophyll cells lose water by evaporation to leaf spaces.
Water vapour diffuses out through open stomata to atmosphere.
Root surface area is greatly increased by presence of root hairs,
    Which form an effective anchorage and surface for the absorption of water.
Absorption of water by root hairs is by osmosis.
An osmotic gradient across the root cortex draws water into the xylem.
In angiosperms xylem carries the water up the stem.
Xylem vessels are open ended, forming long tubes,
    And are also rigid to prevent collapse.
Each column of water is microscopic in diameter.
Adhesive forces cause water to cling to walls of these narrow channels.
Cohesive forces prevent the narrow columns of water snapping.
There must be no air bubbles in columns.
Capillarity due to adhesion and cohesion,
    Is also involved in water uptake.
Force called root pressure is thought to be linked with transpiration,
    Although its role is thought to be minimal.

45. WHAT ADAPTATIONS FOR WATER BALANCE ARE SHOWN BY PLANTS LIVING IN THE FOLLOWING ENVIRONMENTS:

(a) harsh desert conditions;

(b) aquatic conditions?

POINTS TO COVER

(a) Harsh desert conditions

Plants living in harsh desert conditions are called xerophytes.

Guard cells in the leaves may be sunk in pits,
> Water vapour diffusing through open stomata gathers,
> Forming a barrier to prevent further water loss.

Presence of hairs,
> And thickened cuticle help prevent water loss.

Leaves of some xerophytes can roll to cause build-up of humidity and prevent excess water loss,
> Such as marram grass.

Some xerophytes store water in special tissue,
> Called succulent tissue,
> As in cacti.

Plant body shape,
> May present relatively small surface area compared with volume,
> Which helps prevent water loss.

Leaves may be modified as spines to further reduce surface area across which water may be lost,
> As in gorse.

Root system may be specialised,
> Deep or surface spreading.

Plant may show pattern of stomatal opening and closing opposite to that of normal land plants,
> Reversed stomatal rhythm,
> Which means stomata are open at night and closed during the day,
> Conserving water.

(b)   Aquatic conditions

Plants which are adapted to growing submerged (completely or partially),
    Are called hydrophytes.
Hydrophytes may be rooted in water or mud with aerial stems and leaves,
    Or entirely submerged.
Root system is primarily for anchorage only,
    Rather than uptake of minerals, etc.
Common hydrophytes are water lilies, pondweed, etc.
Tissues of hydrophytes are much less compact than normal,
    With many air spaces,
    Which both allow gas exchange with surrounding water,
    And afford support by giving plant body buoyancy.
Usual structural support (xylem tissues, etc.) is missing.
Leaves are extremely delicate,
    Presenting a very large surface area in relation to volume,
    For efficient gas exchange,
    Which occurs directly across the leaf surface,
    Obviating the need for stomata.

46. WHEN RESOURCES BECOME LIMITED, COMPETITION MAY ARISE IN ANIMAL POPULATIONS. BY MEANS OF EXAMPLES, EXPLAIN HOW INTERSPECIFIC AND INTRASPECIFIC COMPETITION CAN OCCUR

POINTS TO COVER

Competition occurs when resources become limited,
    So that more than one organism seeks out those resources at the same time.
Limiting resources might be space for shelter or breeding, food, water, etc.
Particularly evident if habitat in which the organisms live is small.
Competition may be between members of different species,
    Interspecific,
    Or between members of the same species,
    Intraspecific.
Competition has resulted in animals sometimes having sophisticated adaptations to occupy a particular niche.

1. Interspecific Competition

    Generally accepted that if two species occupy the same niche,
        One species will become extinct through interspecific competition,
        The principle of competitive exclusion.
    There is a direct relationship between the degree of similarity of the niches of two species and the probability of interspecific competition developing,
        Hence the observation that adaptive radiation tends to ensure that species occupy a unique niche.

Was shown experimentally using two different species of protozoan *Paramecium caudatum* and *aurelia*.
Similar results have been found in two different species of flour beetle *Tribolium confusum* and *castaneum*.
The findings in such experiments show that when grown separately,
    Each species followed a typical growth pattern,
    However when grown together,
    One species failed to develop (becoming extinct),
    Or developed very slowly.

2. Intraspecific competition

Competition can occur within a species,
  Since individuals are liable to be exploiting the same resources,
  Which may become limiting,
  And thus provoke intraspecific competition.

Individuals have therefore evolved a diverse array of mechanisms,
  To help balance the counter forces of competition and cooperation,
  To reproduce, or avoid predation, etc.

Territorial behaviour is evident in vertebrates (birds, mammals, etc.),
  And some invertebrates (insects, etc.),
  Often functions to prevent another member of the same species,
  From entering a particular area (territory),
  Thus spacing out individuals to avoid intraspecific competition,
  Territories decrease in size as the population size increases,
  Below a certain threshold size,
  Successful reproduction cannot take place.

Establishment of social hierarchy which is particularly evident in vertebrates,
  Based on physical strength, age, health status, etc.,
  Which reduces tension and increases population stability,
  Such as seen in development of pecking order in hens,
  Or group behaviour where one bird acts as a sentinel while other members of the group feed.

47 WHAT ARE THE ADVANTAGES OF SOCIAL BEHAVIOUR?

POINTS TO COVER

Social behaviour has evolved many times in the natural world,
    And is observed in most animals,
    Even those which normally pursue a solitary existence,
    To effect copulation and reproduction for example.

Complex social behaviour is most evident in animals which are naturally social creatures,
    Such as social insects and mammals,
    But is found in other animals also.

Main advantages of living socially include,
    Exploitation of food resources,
        As seen by foraging bees which relay location of food to rest of 'society' via a particular 'dance' pattern,
        And Californian acorn woodpecker which stores acorns for use by all members of the community.
    Better defence of territory,
        As seen in chimpanzees working together in defending an area by attacking intruders.
    Less chance of being taken unaware by predators,
        As seen in sentinel behaviour,
        Of some birds, mongooses and meerkats.
    Increased offensive capability,
        As seen when large numbers of ants together can subdue and kill large prey.
    More efficient use of space,
        As seen in shoal fish maintaining equal spacing between individuals,
        And African bee-eater birds which live in communal burrows carved out by a team of birds.
    Means of communication between members of the same community,
        As seen by elephants raising their ears and trunks which enable individuals to keep in contact with each other, even if separated by several miles.

Allows possible development of hierarchy,
> Which in turn leads to group stability,
> And ensures that in times of shortage,
> The strongest members have access to food, water, shelter, etc.,
> And so have an increased chance of surviving to pass on their genes.

Social behaviour may help prevent aggressive responses between individuals in groups.

Social behaviour has an important genetic basis,
> Since it would take too long for some organisms to acquire such behaviour by learning alone.

48 GIVE AN ACCOUNT OF DIFFERENT WAYS IN WHICH ANIMALS CAN COMMUNICATE WITH EACH OTHER

POINTS TO COVER

Essential in an animal community that individuals be able to communicate with each other.
Even least social animals must communicate,
    To reproduce for example.

Communication may be to,
    Attract a mate,
    Defend territory,
    Effect copulation,
    Obtain food, etc.

Requires a sender of the message (signal),
    As well as a medium to carry message,
        Such as water, air, land,
    And a receiver to pick up the message.

Signals may take many different forms, either singly or collectively.

Visual,
    Such as red-bellied male stickleback reacting aggressively to another rival male,
    Or the striking pattern of many insects to avoid predation,
    Or the pattern of flashing signals produced by fireflies to attract a mate.

Behaviour pattern,
    Such as courtship dances of many tropical birds,
    Or the dancing of bees to indicate source of food,
    Or courtship behaviour of male stickleback in response to female swollen with eggs.

Sound,
    Such as rattle of some snakes to ward off would-be attackers,
    Or the alarm calls in primates,
    Or a hen responding to the distress signals of her chicks,
    Or attraction of a mate by some insects by generating particular frequency using wings,
    Or indication of territorial boundaries by bird song.

Chemical,
>Such as production of noxious chemicals by skin of many amphibia to avoid predation,
>Or the marking-out of territories by dog urination,
>Or the use of chemicals by foraging ants to mark out pathways for other ants,
>Or secretion by moths of chemical attractants which can be detected by a mate over a wide distance,
>>Such chemicals being called pheromones.

Touch,
>Such as grooming in primates,
>>And in prairie dogs.

49 DISCUSS THE EFFECT OF GRAZING BY HERBIVORES WITH REFERENCE TO THE FOLLOWING:

(a) maintenance of species diversity;

(b) ability of plants to tolerate grazing.

POINTS TO COVER

(a) Maintenance of species diversity

Even though plants may have well-developed defences,
  They will still be subject to grazing by a wide range of herbivores,
  Such as ungulates, grasshoppers, rodents, etc.

Evidence of relationship between species diversity and grazing,
  Was seen in the 1950s with the introduction of the myxomatosis virus into Britain to eliminate a rabbit pest problem,
  Which resulted in virtual extermination of this herbivore,
  With a removal of the grazing pressure it had previously exerted,
  Which was very closely followed by an increase in the number of flower species which,
    Either had never previously been found or were present in very low numbers,
  Also followed by an increase in woody plants since seedlings were no longer cropped by rabbits.

Observations on the killing effect of oil spillage along some British coasts,
  On the grazing limpets,
  Which normally live on the rocks,
  And constantly keep the vegetation cover low,
  Now opened up the rock surfaces to a wide range of species of algae which formed a dense covering on these rocks.

The palatability of the vegetation is also an important variable in the effect of grazing on species diversity,
  As seen if sheep graze on land primarily seeded with rye grass and white clover,
    Which are both very palatable to the sheep,
  Grazing is heavy,
  Encouraging the growth of other less palatable vegetation,
  Thus species become diverse as grazing proceeds.

In general the flora is more varied the less selective the grazing herbivore.

(b) Ability of plants to tolerate grazing

Plant tolerance to grazing is usually associated with the ability to grow again and reproduce.

Such tolerance often depends on the anatomy and physiology of the plant,
> So that grasses for example,
>> Which have nodal meristems,
>>> And can therefore tolerate close cropping,
>
> Are much more tolerant than many dicotyledons,
>> Which have apical meristems,
>>> And can therefore not tolerate close cropping.

Grazing can actually enhance primary plant production,
> By preventing vegetation such as grass,
>> Reaching maturity which would then be followed by decline and death.

50. GIVE AN ACCOUNT OF HOW ANIMALS MAY HAVE THEIR BEHAVIOUR MODIFIED UNDER THE FOLLOWING HEADINGS:
(a) habituation;
(b) imprinting.

POINTS TO COVER

(a) Habituation

Response to same stimulus is gradually weakened,
> Provided response is not positively or negatively reinforced by reward or punishment respectively,
> Until it eventually disappears.

Habituation stops useless active responses to stimuli which are unimportant.
Does not involve new response,
> Simply a reduction in the strength of an existing response.

Classical experiments involved withdrawal responses of coelenterates,
> Such as the jellyfish.

Speed of habituation is influenced by interval between repetition of stimulus,
> And by the intensity or type of stimulus.

Widespread in natural world.

(b) Imprinting

Important in early development of animals.
Can be effected by many different stimuli, light, sound, etc.
Easy to demonstrate by experiments.

Imprinting will rapidly influence early behaviour.
Although rapid it is relatively persistent,
> And is very fixed in its response (stereotyped).

Has important survival value,
> Such as chicks who immediately become imprinted with parent.

Lorenz's classical experiments with ducklings, etc.

51 BY MEANS OF SUITABLE EXAMPLES, SHOW HOW PLANTS AND ANIMALS COPE WITH DANGER IN THEIR ENVIRONMENT

POINTS TO COVER

Plants may employ many protective structural and chemical adaptations,
Such as sharp thorns and spines on leaves,
    As seen in holly and cacti.

Or stinging hair cells,
    Which can inject the skin of animals,
    With an irritating cocktail of chemicals,
    As found in nettles.

Or scales to protect delicate tissues,
    From mechanical damage,
    As seen in covering of fleshy bulb parts,
    In daffodils, onions, etc.

Or form a thickened epidermis,
    Which may also be covered with a thick, waxy cuticle,
    Giving some protection from insects.

Or producing increased levels of hormones (such as abscisic acid) during water stress,
    Which restricts plant growth,
    And stimulates stomata in leave to close,
    Thereby conserving water.

Or production of fungicidal chemicals,
    To destroy fungal pathogens at the site of infection.

Or chemicals called 'phytoalexins',
    Which respond to fungal invasion by restricting the growth of the pathogen to healthy tissue.

Animals may employ variety of strategies,
Such as social behavioural adaptations,
    So that even relatively defenceless animals individually,
    Can collectively experience social defence,
    As seen in herds of herbivores,
    Such as zebras, antelopes, etc.,
    Which may graze in herds of upwards of one hundred individuals,
    Making it difficult for a predator,
    Such as a lion,
    To approach undetected one or more of the herd.
    Similar strategy is adopted by many defenceless fish,
    Such as herring,
    Which swim in large shoals,
    Confusing predators,
    Such as barracuda.

Or individual behavioural adaptations,
    Such as habituation and avoidance behaviour,
    In sea-anemones and squids respectively.
    Aggressive displays to ward off possible predators,
    Seen in primates.

Or production of noxious chemicals to ward of predators,
    Found in many amphibia.

Or structural adaptations,
    Such as ability to move quickly to avoid danger,
    As seen in gazelle.